American Folk Tales

美國民間故事

商務印書館

This Chinese edition of *American Folk Tales*
has been published with the written permission of
Black Cat Publishing.

The copyright of this Chinese edition is owned by
The Commercial Press (H.K.) Ltd.

Name of Book: American Folk Tales
Told by: George Gibson
Editors: Monika Marszewska, Elvira Poggi Repetto
Design: Nadia Maestri
Illustrations: Barbara Nascimbeni, Sara Blasigh
Edition: ©1998 Black Cat Publishing
                an imprint of Cideb Editrice, Genoa, Canterbury

系 列 名：Black Cat 優質英語階梯閱讀 · Level 1
書   名：美國民間故事
責任編輯：傅　伊
封面設計：張　毅
出   版：商務印書館（香港）有限公司
         香港筲箕灣耀興道 3 號東滙廣場 8 樓
         http://www.commercialpress.com.hk
發   行：香港聯合書刊物流有限公司
         香港新界大埔汀麗路 36 號中華商務印刷大廈 3 字樓
印   刷：中華商務彩色印刷有限公司
         香港新界大埔汀麗路 36 號中華商務印刷大廈
版   次：2015 年 1 月第 9 次印刷
         © 2003 商務印書館（香港）有限公司
         ISBN 978 962 07 1635 5
         Printed in Hong Kong

出版說明

　　本館一向倡導優質閱讀，近年來連續推出了以 "Q" 為標識的 "Quality English Learning 優質英語學習" 系列，其中《讀名著學英語》叢書，更是香港書展入選好書，讀者反響令人鼓舞。推動社會閱讀風氣，推動英語經典閱讀，藉閱讀拓廣世界視野，提高英語水平，已經成為一種潮流。

　　然良好閱讀習慣的養成非一日之功，大多數初、中級程度的讀者，常視直接閱讀厚重的原著為畏途。如何給年輕的讀者提供切實的指引和幫助，如何既提供優質的學習素材，又提供名師的教學方法，是當下社會關注的重要問題。針對這種情況，本館特別延請香港名校名師，根據多年豐富的教學經驗，精選海外適合初、中級英語程度讀者的優質經典讀物，有系統地出版了這套叢書，名為《Black Cat 優質英語階梯閱讀》。

　　《Black Cat 優質英語階梯閱讀》體現了香港名校名師堅持經典學習的教學理念，以及多年行之有效的學習方法。既有經過改寫和縮寫的經典名著，又有富創意的現代作品；既有精心設計的聽、說、讀、寫綜合練習，又有豐富的歷史文化知識；既有彩色插圖、繪圖和照片，又有英美專業演員朗讀作品的 CD。適合口味不同的讀者享受閱讀之樂，欣賞經典之美。

　　《Black Cat 優質英語階梯閱讀》由淺入深，逐階提升，好像參與一個尋寶遊戲，入門並不難，但要真正尋得寶藏，需要投入，更需要堅持。只有置身其中的人，才能體味純正英語的魅力，領略得到尋寶的快樂。當英語閱讀成為自己生活的一部分，英語水平的提高自然水到渠成。

商務印書館（香港）有限公司
編輯部

使用說明

1 應該怎樣選書?

按閱讀興趣選書

《Black Cat 優質英語階梯閱讀》精選世界經典作品,也包括富於創意的現代作品;既有膾炙人口的小說、戲劇,又有非小說類的文化知識讀物,品種豐富,內容多樣,適合口味不同的讀者挑選自己感興趣的書,享受閱讀的樂趣。

按英語程度選書

《Black Cat 優質英語階梯閱讀》現設 Level 1 至 Level 6,由淺入深,涵蓋初、中級英語程度。讀物分級採用了國際上通用的劃分標準,主要以詞彙 (vocabulary) 和結構 (structures) 劃分。

Level 1 至 Level 3 出現的詞彙較淺顯,相對深的核心詞彙均配上中文解釋,節省讀者查找詞典的時間,以專心理解正文內容。在註釋的幫助下,讀者若能流暢地閱讀正文內容,就不用擔心這一本書程度過深。

Level 1 至 Level 3 出現的動詞時態形式和句子結構比較簡單。動詞時態形式以現在時 (present simple)、現在時進行式 (present continuous)、過去時 (past simple) 為主,句子結構大部分是簡單句 (simple sentences)。此外,還包括比較級和最高級 (comparative and superlative forms)、可數和不可數名詞 (countable and uncountable nouns) 以及冠詞 (articles) 等語法知識點。

Level 4 至 Level 6 出現的動詞時態形式,以現在完成時 (present perfect)、現在完成時進行式 (present perfect continuous)、過去完成時 (past perfect continuous) 為主,句子結構大部分是複合句 (compound sentences)、條件從句 (1st and 2nd conditional sentences) 等。此外,還包括情態動詞 (modal verbs)、被動形式 (passive forms)、動名詞 (gerunds)、

短語動詞（phrasal verbs）等語法知識點。

　　根據上述的語法範圍，讀者可按自己實際的英語水平，如詞彙量、語法知識、理解能力、閱讀能力等自主選擇，不再受制於學校年級劃分或學歷高低的約束，完全根據個人需要選擇合適的讀物。

❷ 怎樣提高閱讀效果？

　　閱讀的方法主要有兩種：一是泛讀，二是精讀。兩者各有功能，適當地結合使用，相輔相成，有事半功倍之效。

　　泛讀，指閱讀大量適合自己程度（可稍淺，但不能過深）、不同內容、風格、體裁的讀物，但求明白內容大意，不用花費太多時間鑽研細節，主要作用是多接觸英語，減輕對它的生疏感，鞏固以前所學過的英語，讓腦子在潛意識中吸收詞彙用法、語法結構等。

　　精讀，指小心認真地閱讀內容精彩、組織有條理、遣詞造句又正確的作品，着重點在於理解 "準確" 及 "深入"，欣賞其精彩獨到之處。精讀時，可充分利用書中精心設計的練習，學習掌握有用的英語詞彙和語法知識。精讀後，可再花十分鐘朗讀其中一小段有趣的文字，邊唸邊細心領會文字的結構和意思。

　　《Black Cat 優質英語階梯閱讀》中的作品均值得精讀，如時間有限，不妨嘗試每兩個星期泛讀一本，輔以每星期挑選書中一章精彩的文字精讀。要學好英語，持之以恆地泛讀和精讀英文是最有效的方法。

❸ 本系列的練習與測試有何功能？

　　《Black Cat 優質英語階梯閱讀》特別注重練習的設計，為讀者考慮周到，切合實用需求，學習功能強。每章後均配有訓練聽、說、讀、寫四項技能的練習，分量、難度恰到好處。

聽力練習分兩類，一是重聽故事回答問題，二是聆聽主角對話、書信朗讀、或模擬記者訪問後寫出答案，旨在以生活化的練習形式逐步提高聽力。每本書均配有CD提供作品朗讀，朗讀者都是專業演員，英國作品由英國演員錄音，美國作品由美國演員錄音，務求增加聆聽的真實感和感染力。多聆聽英式和美式英語兩種發音，可讓讀者熟悉二者的差異，逐漸培養分辨英美發音的能力，提高聆聽理解的準確度。此外，模仿錄音朗讀故事或模仿主人翁在戲劇中的對白，都是訓練口語能力的好方法。

閱讀理解練習形式多樣化，有縱橫字謎、配對、填空、字句重組等等，注重訓練讀者的理解、推敲和聯想等多種閱讀技能。

寫作練習尤具新意，教讀者使用網式圖示（spidergrams）記錄重點，採用問答、書信、電報、記者採訪等多樣化形式，鼓勵讀者動手寫作。

書後更設有升級測試（Exit Test）及答案，供讀者檢查學習效果。充分利用書中的練習和測試，可全面提升聽、說、讀、寫四項技能。

◆4 本系列還能提供甚麼幫助？

《Black Cat 優質英語階梯閱讀》提倡豐富多元的現代閱讀，巧用書中提供的資訊，有助於提升英語理解力，擴闊視野。

每本書都設有專章介紹相關的歷史文化知識，經典名著更有作者生平、社會背景等資訊。書內富有表現力的彩色插圖、繪圖和照片，使閱讀充滿趣味，部分加上如何解讀古典名畫的指導，增長見識。有的書還提供一些與主題相關的網址，比如關於不同國家的節慶源流的網址，讓讀者多利用網上資源增進知識。

Contents

The Tale of Brer Rabbit and the Tar Baby
兔兄弟和柏油娃娃的故事

The story is recorded in full. 故事錄音

This symbol indicates the exercises featured on the accompanying CD. 聽力練習的錄音標記

The Legend of Johnny Appleseed

1 **Do you know these words?**

fence
籬笆

seeds
種子

log cabin
小木屋

orchard
果園

military fort
軍事要塞

2 **Match these definitions with the words in the red apple.**

a. A big bag to carry things is called a
....

b. .... go to another land to live
there permanently.

c. Books are kept in a ....

d. You .... your leg when you
have an accident and fall.

e. A .... takes a message
from one place to another.

library   sack
settlers
messenger
injure

3 Johnny began his journey in Boston. He walked to New York, Pennsylvania and to the Midwest. Look at the map. Did Johnny walk to the north, south, east or west?

Johnny and his Apple Seeds

Johnny Appleseed was born near Boston in 1775. His real name was John Chapman.

When he was a child he played in the forest and in the fields. His best friends were animals. He loved all animals. He played with them and talked to them! His family was very religious [1]. Johnny's first book was the Bible, but he also liked *Aesop's Fables*. Johnny loved the tales about animals and their adventures.

When Johnny was a teenager he worked as a missionary [2] with the Indians. He converted [3] many Indians to Christianity [4]. He taught them about the Bible. The Indians were his friends.

When he was 26 years old he had a vision [5]. An angel appeared to

---

1. **religious** : 篤信宗教的，虔誠的。
2. **missionary** : 傳教士。
3. **converted** : 轉變，轉換。
4. **Christianity** : 基督教。
5. **vision** : 幻想，異象。

him! The angel said, "Go and plant apple seeds across America. The settlers of the new frontier want good apples to eat."

Johnny was surprised but he was happy. He was a kind person and he wanted to help others. He took a big sack[1] and filled it with apple seeds. He carried this sack on his back. In one hand he carried the Bible, *Aesop's Fables* and other religious books. Now he was ready to cross the continent and plant America's favourite fruit: the apple.

Johnny was an unusual[2] man. He was tall and thin. He had long hair and a beard. He never bought new clothes. He wore an old coffee sack and the old clothes people gave him. He didn't usually wear any shoes. He wore a saucepan[3] on his head.

One of Johnny's friends said, "God bless you, Johnny. We are happy for you. You are similar to[4] St. Francis of Assisi. He loved animals and lived a simple life."

Johnny said, "I want to plant apple seeds across America. Every American family will have apple trees with good apples to eat."

In 1800 Johnny began his long journey across America. At that time America was a very young country. The American continent was a wilderness[5]. It was unexplored[6]. There were no roads and

---

1. **sack**：大口袋。
2. **unusual**：異乎尋常的。
3. **saucepan**：
4. **similar to**：與⋯相似的。
5. **wilderness**：荒野。
6. **unexplored**：未經勘察的。

The Legend of Johnny Appleseed

few maps. This immense [1] land was called the American frontier. Many settlers wanted to explore the frontier.

Johnny walked from Massachusetts to New York. From New York he walked to Pennsylvania. Then he crossed Ohio, Indiana and a big part of the Midwest (see map on page 11). Every day he moved west.

He travelled across America and planted apple seeds. He built fences around the fields and then continued his journey. Settlers travelled to the frontier and found the apple orchards. They ate the delicious [2] fruit: green, red and yellow apples. When the settlers found an apple orchard, they built a home there.

Other settlers dug up [3] the apple trees and took them to new lands. Some of Johnny's trees travelled to the West Coast on the Pacific Ocean [4]. (see map on page 11)

When Johnny found a family of settlers he visited their log cabin. He helped them with their work. He told the children stories and sang songs.

One day Johnny visited a family of settlers in the Midwest. This family loved books. He gave them a few pages from his books. "You can read them and give them to me when I return in a few months," he said. The family was very happy.

In this way, Johnny created the first library on the frontier. Many children learned to read thanks to Johnny and his library.

1. **immense**：極大的。
2. **delicious**：美味的。
3. **dug up**：(dug 是動詞 dig 的過去時) 從地下挖掘出某物。
4. **the Pacific Ocean**：太平洋。

UNDERSTANDING THE TEXT

1 **Choose the correct answer.**

a. Johnny Appleseed's family was very religious and
- ☐ Johnny worked as a missionary with the Indians
- ☐ Johnny read the Bible every day
- ☐ Johnny went to Asia to work as a missionary

b. An angel appeared to Johnny and said
- ☐ "Go and teach the Bible to the Indians"
- ☐ "Go and plant apple seeds across America"
- ☐ "Go and sell apples across America"

c. Johnny filled a big sack with
- ☐ books and a saucepan
- ☐ apple seeds
- ☐ old clothes and *Aesop's Fables*

d. Johnny walked from Massachusetts to
- ☐ the West Coast
- ☐ Canada
- ☐ the Midwest

e. Johnny built fences
- ☐ around the fields
- ☐ in Massachusetts
- ☐ on the West Coast

f. Settlers travelled to the West and found
- ☐ apple seeds
- ☐ religious books
- ☐ apple orchards

g. Johnny Appleseed created
- ☐ roads and maps
- ☐ the first library on the frontier
- ☐ a new religion

2 **Look at these sentences from Part One:**

Johnny was an *unusual* man.　　　(*unusual=not usual*)
It was *unexplored*.　　　　　　　(*unexplored=not explored*)

The prefix（前綴）"-un" before a word often means *not*.
Look at this example:

The trip is **unnecessary**. = The trip is **not necessary**.

Use the prefix "-un" to create new words from the ones given in the sentences.

a. The soldier was not kind to the animals.

....

b. Johnny was not able to swim.

....

c. The Indians were not happy when Johnny left.

....

d. The children had little food and were not healthy.

....

e. The house was not tidy.

....

f. That language is not known.

....

3 **Look at these words and then find their opposites in Part One. One is done for you.**

a. sad ....happy............................

b. small ....

c. fat ....

d. short ....

e. new ....

f. bad ....

g. last ....

4 **Use some of your answers to describe Johnny Appleseed.**

Johnny was a kind person and he wanted to help others. He was tall and .... He had .... hair and a beard. He wore an .... coffee sack. He didn't usually wear any shoes.

He loved animals and lived a simple life. He created the .... library on the frontier.

5 **Here you have the Infinitives of the irregular verbs** (不規則動詞的原形) **in Part One. Go back to Part One and underline the Past Simple of the verbs** (動詞過去時) **below, then write them next to their Infinitives.**

Infinitive	Past Simple	Infinitive	Past Simple
teach	....	give	....
have	....	begin	....
say	....	build	....
take	....	find	....
buy	....	eat	....
wear	....	tell	....

6 Listen to the first four paragraphs of Part One and put the pictures in the order that they are mentioned. Write 1, 2, 3 etc. in the correct box.

AMERICAN FOOD
made with Apples

Americans eat a lot of green, yellow and red apples. There is a lot of food and drink made with apples. Here is a proverb [1]: "An apple a day keeps the doctor away."

Apples are a delicious fruit and are very good for you. Today in the United States there are more than 7,000 different kinds of apples!

Look at this food made with apples:

apple tart [2]

apple cake

baked apple

apple bread

apple cider [3]

apple pie

1. **proverb** : 諺語。　　2. **tart** : 果餡餅。　　3. **cider** : 蘋果酒。

At Halloween, children in America "bob for apples [1]".

1 What food is made with apples in your country?
What's your favourite fruit?
What fruits have Vitamin C?

PART TWO

Johnny becomes a Legend

For many years Johnny walked thousands of miles across the frontier. He planted apple seeds, lent [1] books, protected animals and made friends with settlers and Indians.

The Indians liked Johnny because he had no weapons [2] and respected [3] nature. Johnny learned to speak the languages of many tribes.

One summer morning many Indians arrived at a frontier village. The Indians wanted to destroy the village and kill the settlers. The settlers wanted to send a messenger to a military fort to ask for help. The fort was 30 miles away. It was very dangerous [4]. There were enemy Indians everywhere.

---

1. **lent** :（動詞 lend 的過去時）把某物借給別人。

2. **weapons** : 武器。

3. **respected** : 尊敬。

4. **dangerous** : 有危險的。

American Folk Tales

Johnny wanted to save the lives of the settlers. He knew all the secret paths [1] in the forest and in the mountains. He was not afraid.

He took the message to the fort. The soldiers at the fort got on their horses and went to defend the settlers.

A few years later, Johnny visited an Indian village. He heard the Indian chief say, "Many settlers live near the river. Tonight we will kill all of them."

That evening Johnny ran to the home of every settler and said, "The Indians will attack you tonight. Run away! Go and hide in the forest!" The settlers escaped [2] to the forest and no one was killed.

Johnny loved all forms of human and animal life. He didn't eat meat because he didn't want to kill animals. He loved insects, too. He loved and respected every living thing. He was a very special person.

One autumn day, Johnny was near an apple orchard. He heard the cry of an animal. Behind a tree he found a deer [3].

"My poor friend!" said Johnny. "Don't be afraid!"

He examined the deer and said, "Your leg is injured [4]. I can help you." The deer wasn't afraid.

Johnny stayed with the deer for many days and helped it.

"I'm happy you can walk again. Go and run in the forest, my little friend!"

---

1. **paths** : 小路。

2. **escaped** : 逃脱。

3. **deer** :

4. **injured** : 受傷的。

American  Folk Tales

During a snow storm, Johnny wanted to sleep inside a small cave. He saw a big family of racoons[1] sleeping inside the cave. He did not want the racoons to go outside into the snow. So Johnny slept outside in the cold!

For almost 50 years, Johnny Appleseed helped the American frontier to grow. His apple trees, his books, his generosity[2] and his kindness made the frontier a happy place.

Johnny became a legend[3] during his life. Everyone loved him. He was a true friend of the settlers, the Indians and the animals.

In America today, people remember him with admiration[4]. When Americans eat an apple, they often think of Johnny Appleseed who made apple trees grow all over America.

1. **racoons :**
2. **generosity :** 慷慨。
3. **legend :** 傳奇。
4. **admiration :** 讚賞。

UNDERSTANDING THE TEXT

1 Complete the following sentences with the words in the tree.

message

forest home happy

thousands animal

apple seeds fort

attack legend village

everyone

Johnny walked .... of miles across the frontier. He made friends with .... He planted .... and lent books.

One summer morning, the Indians wanted to destroy a frontier .... Johnny walked 30 miles to take a .... to a .... He saved the lives of the settlers.

Another time he ran to the .... of every settler and said, "The Indians will .... you tonight. Run away! Go and hide in the .... !"

Johnny loved all forms of human and .... life. He made the frontier a .... place. He became a .... during his life.

In general we use *some* **in positive sentences**（肯定句）**and** *any* **in negative and interrogative sentences**（否定句和疑問句）**.**
Look at these examples:

Johnny gave **some** apple seeds to the settlers.
He didn't have **any** money.
Did you see **any** Indians?

Fill in the gaps with *some* **or** *any.*

a. There were .... racoons in the cave.

b. Does he have .... seeds in his sack?

c. He didn't plant .... apple seeds in the forest.

d. The settlers wanted .... books to read.

e. They gave us .... fruit to eat.

f. Did the soldiers find .... settlers in the mountains?

g. We didn't find .... people in the fort.

3 **Have fun with this crossword puzzle!**

ACROSS

1.

2.

3.

4. Johnny became a ....

5.

DOWN

6. Johnny's first book.

7.

8.

9.

10. Johnny wanted to sleep here.

4 You are a journalist. You work for the *Boston Gazette*. In Boston everyone wants to read about Johnny Appleseed and his adventures on the frontier. You must write a short article about Johnny.
You received this telegram from a fort in the Midwest but some words are scrambled. Unscramble the words, write them in the spaces provided and your newspaper article is ready!

Western Union Telegram

A *(verba)* .... young man saved *(elrsetst)*
.... from Indian attack and fire. The
young man's *(eanm)* .... is Johnny
Appleseed.

He went through *(yemen)* .... territories.
He *(koto)* .... a message to a *(roft)*
....

The *(lssrdieo)* .... at the fort got on
their horses and went to *(aves)* .... the
settlers.

Johnny is a *(inferd)* .... of the Indians,
the settlers and the *(mlasnai)* ....

He is a *(gnlede)* ....

YOUNG AMERICA
and its Settlers

When Johnny Appleseed was born in 1775, America was not a nation. It was an English colony. America became a nation after the American Revolution.

The enormous [1] American continent was a wilderness. There were unexplored forests, mountains, rivers, lakes, fields and deserts. Only the American Indians lived in small parts of this wilderness.

The American Army [2] sent its soldiers to build forts in the wilderness. These forts were similar to small villages. They protected settlers from Indian attacks. They also

1. **enormous**：巨大的。
2. **Army**：軍隊。

sold supplies [1]. Settlers often built their log cabins near military forts.

Thousands of settlers wanted to live in these new lands. It was their dream. The settlers didn't usually buy the land. It was free from the Government.

Families of settlers travelled in covered wagons. These covered wagon trains travelled long distances to new lands. Settlers were strong, courageous people. Life on the frontier was not easy. It was difficult and dangerous.

The men built log cabins and hunted [2] for food. The women and children worked as farmers. There was little free time.

The settlers were happy. They loved the freedom and adventure of the frontier. They built America!

A covered wagon train

1. **supplies** : 供給。
2. **hunted** : 打獵。

1 Choose the correct answer.

a. America became a nation
- [ ] before the American Revolution.
- [ ] when Johnny Appleseed was born.
- [ ] after the American Revolution.

b. American Indians lived
- [ ] everywhere in the wilderness.
- [ ] in small parts of the wilderness.
- [ ] in log cabins in the wilderness.

c. The American Army built forts
- [ ] to protect settlers and sell supplies.
- [ ] to explore the West.
- [ ] to attack the Indians.

d. Families of settlers
- [ ] bought land.
- [ ] travelled in covered wagons.
- [ ] lived in forts.

e. The men
- [ ] built log cabins and hunted for food.
- [ ] worked in the forts.
- [ ] built covered wagons.

f. Life on the frontier was dangerous,
- [ ] but the settlers found a lot of gold.
- [ ] and most settlers died.
- [ ] but the settlers loved the freedom and adventure.

Inside a log cabin

An Indian attack

2 **How many words can you make from this title?**

THE AMERICAN FRONTIER

Two are done for you.

can		
fire		

3 **Listen to the first three paragraphs of "Young America and its Settlers". Then listen again and fill in the gaps.**

When Johnny Appleseed was .... in 1775, America was not a nation. It was an .... colony. America became a nation .... the American Revolution.

The enormous .... continent was a wilderness. There were unexplored ...., mountains, rivers, lakes, fields and .... Only the American Indians lived in .... parts of this wilderness.

The American Army sent its soldiers to build forts in the wilderness. These .... were similar to small villages. They protected settlers from Indian attacks. They also .... supplies. Settlers often built their .... cabins near military forts.

Pecos Bill

BEFORE READING

 Do you know these words?

covered wagon
大篷車

cactus (plural=cacti)
仙人掌

coyote
郊狼

ranch
大牧場

plains
平原

cattle
牛

rope
繩子

blackbird
黑鸝

PART ONE

Bill and the Coyotes

Pecos Bill was born in the East of the United States in the 1800's. He had a lot of brothers and sisters. Bill was the baby of the family.

When Bill was about two years old, his mother and father decided to move to the West. They wanted to be pioneers [1]. They liked the adventure of the frontier.

One day they put all their possessions [2] in a covered wagon and began their journey. The family crossed forests, mountains, rivers and plains. They saw many new things. They met friendly Indians.

When they arrived in Texas, little Bill fell out of the covered wagon! His brothers and sisters didn't see him fall out. His parents didn't see him fall out! That evening his parents looked for him. They looked everywhere, but they did not find little Bill. They were very sad but they continued their journey.

---

1. **pioneers** : 拓荒者。
2. **possessions** : 所有物，財產。

American Folk Tales

Little Bill was all alone in the plains of South Texas. He was an intelligent [1] child. He looked around. He saw mountains, cacti and other small plants. It was very hot. Then he saw a cave. He went inside the cave and slept. He slept for a long time.

In the cave there was a family of coyotes. There was a mother coyote and five small coyotes. The mother coyote liked little Bill. She decided to protect him. Bill liked his new mother and his new brothers and sisters. He played with the little coyotes. The coyote family was kind to him. He copied the coyotes and learned to run and eat. He learned to drink water from the river.

At night Bill howled [2] at the moon with the coyotes. Bill learned to speak the language of the animals. Soon Bill forgot about his human family. He thought he was a coyote!

Many years passed and Bill grew up. One day when he was about 20 years old, Bill was at the Pecos River with the other coyotes. He drank water with them. A cowboy saw him and asked, "Why are you drinking water in that way? You aren't a coyote!"

Bill looked at the cowboy and said, "Yes, I am!"

"No! You're not a coyote!" said the cowboy.

"Of course I'm a coyote! This is my coyote family," said Bill.

"Where's your tail [3]?" asked the cowboy.

---

1. **intelligent**：聰明的。
2. **howled**：嗥叫。

3. **tail**：

American Folk Tales

Bill looked in the water of the Pecos River to see his reflection [1]. He didn't see his tail. He looked and looked. He turned around and looked again. He was surprised. He didn't have a tail!

"You're right. I'm not a coyote. My name is Bill. I'm a human! This is a big surprise for me," he said.

The cowboy laughed and Bill laughed too.

"My name's Tall Tom. I'm a very tall cowboy. This is the Pecos River and I will call you Pecos Bill! Come with me, Pecos Bill! You can be a cowboy with me. We can work together at the Longhorn Ranch."

"What's a cowboy?" asked Pecos Bill.

"A cowboy watches and guards the cattle and horses. He takes cattle from Texas to other places. People from the East buy our cattle. They like good meat. A cowboy is a strong, courageous man."

Pecos Bill wanted to be a cowboy. He said goodbye to the coyotes. He was sad to leave them. They were his friends.

Then he said to Tall Tom, "Let's go to the ranch!"

The cowboy gave Pecos Bill some cowboy clothes. Pecos Bill looked at the clothes and laughed. Then he put them on.

Tall Tom rode his horse. Pecos Bill walked near him because he didn't have a horse. It was a sunny day. The sky was blue and the sun was hot.

---

1. **reflection**：倒影。

UNDERSTANDING THE TEXT

1 Fill in the gaps with the correct words from the cactus. Some words can be used more than once.

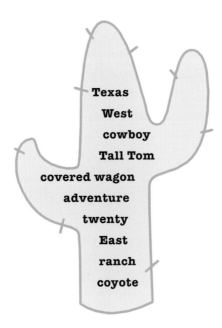

Texas
West
cowboy
Tall Tom
covered wagon
adventure
twenty
East
ranch
coyote

a. Pecos Bill was born in the .... of the United States.

b. His mother and father liked the .... of the frontier. They decided to move to the ....

c. When the family arrived in ...., little Bill fell out of the ....

d. A mother .... liked Bill and decided to protect him.

e. When Bill was about .... years old he met a .... The cowboy's name was ....

f. Bill decided to become a .... and work on a ....

Find the hidden word

Take the letters you need to make the words that match the pictures. Put the remaining letter in the box below. What word do you get?

a. tcsucta

b. iearvr

c. iocyeot

d. eorfslt

Bill didn't have a [a.] [b.] [c.] [d.] .

3 **Complete the sentences below. Use the Past Simple tense of the infinitives**（動詞過去時）**in the covered wagon.**

be   forget   put

sleep   go   fall

walk   decide

begin   ride

a. When Bill .... two years old, his family
.... to move to the West.

b. One day they .... all their things in a covered
wagon and .... their journey.

c. Little Bill .... out of the covered wagon.

d. Bill .... inside the cave and ....

e. He soon .... about his human family.

f. Tall Tom .... his horse.

g. Bill .... near Tall Tom.

4 **What's the weather like?**

Look at these pictures and describe the weather. Use the words in the cowboy hat to help you.

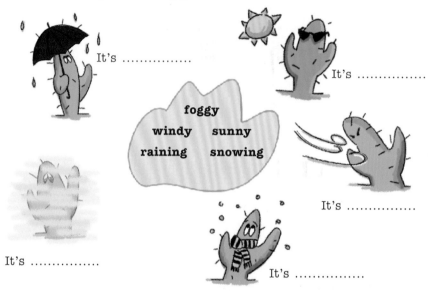

It's ....

It's ....

foggy
windy   sunny
raining   snowing

It's ....

It's ....

It's ....

5 **Listen to the first three paragraphs of Part One and fill in the missing words. If necessary, listen to the text twice.**

Pecos Bill was .... in the East of the United States in the 1800's. He had a .... of brothers and sisters. Bill was the baby of the family.

When Bill was about two years old, his mother and .... decided to move to the .... They wanted to be pioneers. They liked the adventure of the frontier.

One day they put all their possessions in a covered .... and began their journey. The family crossed ...., mountains, rivers and plains. They .... many new things. They .... friendly Indians.

46

PART TWO

King of the Cowboys

Pecos Bill and Tall Tom travelled under the hot sun. It was a long journey. When they arrived at the mountains, it was evening. Pecos Bill said, "I'm tired. Let's sit down and rest."

"That's a good idea," said Tall Tom. "My horse is tired and thirsty [1]." He gave his horse some water to drink.

Pecos Bill and Tall Tom sat down under a big tree. There was a blackbird in the tree. It sang a happy song. Pecos Bill knew the language of the animals so he spoke to the blackbird. They had a long conversation [2].

Pecos Bill and Tall Tom ate some biscuits and drank some water. Pecos Bill gave the blackbird some of his biscuit.

1. **thirsty** : 渴的。
2. **coversation** : 交談。

American  Folk Tales

Tall Tom sang a Western song. Pecos Bill liked it and asked, "What's the name of the song?"

"It's called *Red River Valley*. Many cowboys in Texas sing it," said Tall Tom.

Tall Tom made a fire and they talked about a cowboy's life. They looked at the stars in the night sky. Then they fell asleep. Early the next morning they began their journey to the ranch.

After two days they finally arrived at the cattle ranch. The other cowboys were happy to meet Pecos Bill.

"Welcome to Longhorn Ranch," said the cowboys.

The ranch was very big. There were longhorn cattle [1] everywhere. They ate grass and drank water at the river.

Pecos Bill saw the cowboys on their horses. Every cowboy had a long rope in his hands.

"I want to be a cowboy," Pecos Bill said to Tall Tom. "What must I do?"

"First, you must find a horse. Second, you must have a rope. Then we must take all the cattle to the Red River Valley. We must sell the cattle there. The Red River Valley is far away. It is a long, difficult journey."

Pecos Bill looked around. He saw a black horse near a cactus. No one wanted to ride that horse. He went to the horse and talked to it in animal language. The horse didn't answer. Pecos Bill got on the horse. He tried to ride it.

After a few moments the black horse bucked him off [2]! Pecos Bill tried again. The black horse bucked him off again.

---

1. **longhorn cattle**：見第57頁照片。
2. **bucked him off!**：把他摔下。

King of the Cowboys

"You're a bucking bronco [1]! That's your new name: Bucking Bronco," said Pecos Bill to the horse.

Bucking Bronco bucked Pecos Bill off for three days. He did not want a master. Pecos Bill was strong and determined [2]. He wasn't afraid of Bucking Bronco. He wasn't afraid of anything!

Bucking Bronco was a beautiful horse, but he was very wild. On the fourth day Bucking Bronco stopped bucking. He understood that Pecos Bill was a special cowboy. Pecos Bill was very strong. Bucking Bronco knew that Pecos Bill was his new master.

1. **bucking bronco**：摔人的野馬。
2. **determined**：堅定的。

American   Folk Tales

Pecos Bill and Bucking Bronco became good friends. Together they roped [1] the cattle of Texas. Together they took cattle from Texas to other states.

Pecos Bill became a famous cowboy. He was the best cowboy at the rodeos [2]. Everyone knew him and liked him. He became the King of the Cowboys of Texas!

After a few years, Pecos Bill met a beautiful woman called Sue. She was very friendly. Her home was near the Rio Grande River. Pecos Bill loved her and she loved him.

One sunny day in April they got married. Everyone at the Longhorn Ranch celebrated [3] the wedding. There was an exciting rodeo. There was music, dancing and a lot of good food.

Soup Sam, the friendly cook, organised a pie [4] -eating competition.

Tall Tom's favourite food was apple pie. But he did not win the competition. His friend Big Bob won the pie-eating competition. Big Bob ate 88 apple pies!

Pecos Bill and Sue were very happy at the Longhorn Ranch. Pecos Bill never forgot his friends the coyotes. And he never forgot the language of the animals.

---

1. **roped**：用繩捆綁。
2. **rodeos**：牛仔騎術表演或競賽。

3. **celebrated**：慶祝。

4. **pie**：

UNDERSTANDING THE TEXT

Choose the correct answer.

a. When Pecos Bill and Tall Tom arrived at the mountains
☐ they met a mother coyote.
☐ they met some Indians.
☐ they sat down to rest.

b. At the Longhorn Ranch the cowboys
☐ were happy to meet Pecos Bill.
☐ had dinner.
☐ slept.

c. Pecos Bill called his new horse
☐ Beautiful Buck.
☐ Bucking Bronco.
☐ Black Beauty.

d. Pecos Bill and his horse took the cattle
☐ to the mountains.
☐ to the rodeo.
☐ from Texas to other states.

e. Pecos Bill became
☐ the King of the Cowboys of Texas.
☐ the King of Texas.
☐ the King of Longhorn Ranch.

f. Who won the pie-eating competition?
☐ Soup Sam.
☐ Big Bob.
☐ Tall Tom.

2 This is a newspaper article from the *Texas Star*. It is an article about Pecos Bill. The writer of the article forgot to use the genitive（所有格）"s".
Can you help the writer? Look at the highlighted sections, use the genitive "s" and then copy the article in your notebook. Look at the first example.

Texas has a King!

Pecos Bill's success
The success of Pecos Bill is everywhere. He is the King of the Cowboys of Texas.

Pecos Bill was born in the East. *The family of Pecos Bill* wanted to move to the West. *The dream of the family* was to go to California.

During the journey, little Bill fell out of *the covered wagon of the family*.

A friendly coyote looked after little Bill. One day a cowboy met him and took him to Longhorn Ranch.

The horse of Pecos Bill is called Bucking Bronco. *The best friend of Pecos Bill* is Tall Tom. *The favourite food of Tall Tom* is apple pie.

Last month Pecos Bill married Sue, a beautiful woman. *The home of Sue* is near the Rio Grande River. *The wedding celebration of Pecos Bill* was exciting. Pecos Bill and Sue now live at the Longhorn Ranch.

3 **Word search**

Find four names of animals that you read about in the story and circle them.

S	U	B	J	K	E	H	Y	F	X	V	P	O
R	E	S	Z	T	B	C	O	C	L	B	N	G
A	O	C	K	C	L	L	A	T	G	A	U	K
T	G	O	U	V	A	E	N	T	D	N	T	Z
O	D	Y	A	O	C	G	K	V	T	N	E	T
F	U	O	E	Q	K	S	U	K	G	L	A	R
V	E	T	K	C	B	D	H	O	R	S	E	K
X	N	E	V	M	I	J	G	E	S	Z	M	Q
P	O	C	H	D	R	P	V	E	R	A	U	S
V	G	P	B	K	D	O	F	Z	B	N	P	G

Now use the words to complete these sentences.

a. Bill thought he was a ....

b. Pecos Bill and a .... had a long conversation.

c. The cowboys took the .... to the Red River Valley.

d. Pecos Bill rode a beautiful black ....

53

4 **Who are they?**

Read the clues and match the descriptions with the correct names.

Who...

1. ate 88 apple pies?
2. gave Pecos Bill cowboy clothes?
3. looked after little Bill?
4. did Pecos Bill marry?
5. was the cook at the Longhorn Ranch?
6. was Pecos Bill's horse?

a. ☐ Big Bob
b. ☐ a mother coyote
c. ☐ Bucking Bronco
d. ☐ Soup Sam
e. ☐ Sue
f. ☐ Tall Tom

5 **Word pyramid**

Find the missing words and build the word pyramid.

a. After two days they arrived _ _ the ranch.
b. It was a _ _ _ _ journey.
c. There were longhorn _ _ _ _ _ _ everywhere.
d. Bill talked to the horse in animal _ _ _ _ _ _ _ _ .
e. Everyone at the ranch _ _ _ _ _ _ _ _ _ _ the wedding.

a. ☐☐
b. ☐☐☐☐
c. ☐☐☐☐☐☐
d. ☐☐☐☐☐☐☐
e. ☐☐☐☐☐☐☐☐☐

6 This is a map of the United States. Look at Pecos Bill's journey and say what happened. Fill in the gaps with the correct words in the horse.

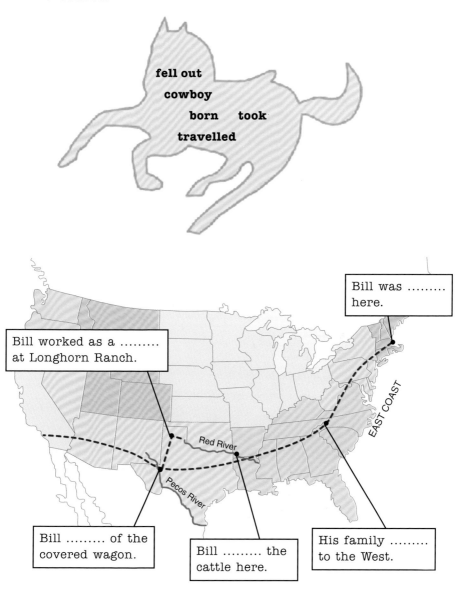

fell out
cowboy
born   took
travelled

Bill was .... here.

Bill worked as a .... at Longhorn Ranch.

EAST COAST

Red River

Pecos River

Bill .... of the covered wagon.

Bill .... the cattle here.

His family .... to the West.

55

Life on a Ranch

Cowboys usually lived and worked on a ranch. A ranch was a very big piece of land. There was usually only one owner [1] of a ranch. There were cattle, sheep and horses on a ranch. Rodeos were a favourite pastime [2]. When cowboys were at the ranch, they lived together in a bunkhouse [3]. The owner of the ranch lived in another house.

A rodeo

Cowboys at work

1. **owner**：所有者。
2. **pastime**：消遣。
3. **bunkhouse**：牛仔的臨時住房。

The cook

The cowboys ate together in a big room. Cowboys had big appetites [1]! The cook was an important person.

Cowboys often moved cattle to another place to sell them. Ten or twelve cowboys moved about 3,000 cattle!

Longhorn cattle

1. **appetites**：胃口。

This was difficult work. They usually travelled for many weeks. They lived on the plains and in the mountains. They cooked their meals on an open fire [1]. They slept under the stars. Sometimes it was very hot and other times it rained or snowed. Sometimes during the long journey there were Indian attacks.

When the cowboys sold the cattle they were very happy. They stayed in a town for a few days. They bought new things. They went to the town saloon [2]. At the saloon they played cards, drank whisky and had a bath! Then they returned to their ranch.

Today there are many cattle ranches in the United States. These ranches are in Texas and in the Western states. Cowboys work on these ranches and they still ride horses!

A cowboy and his horse

1. **open fire :** 　　　　2. **saloon :** 酒館。

1 **Are these sentences true (T) or false (F)? Correct the false sentences.**

	T	F
a. A ranch is a very big building.	☐	☐
b. The owner of the ranch lived in a bunkhouse.	☐	☐
c. Cowboys had big appetites.	☐	☐
d. The cowboys' work was easy.	☐	☐
e. Cowboys usually travelled for many weeks.	☐	☐
f. They stayed in a town for a few hours.	☐	☐
g. Today cowboys work on ranches and they still ride horses.	☐	☐

A cowboy moving cattle

2 **What did the cowboys do in these places? The answers are in the open fire.**

At the Ranch

a. They .... and ....

b. They .... together in a big room.

At the Saloon

a. They .... cards.

b. They .... whisky and .... a bath.

On the Plains and in the Mountains

a. They .... their meals on an open fire.

b. They .... under the stars.

worked   played
ate   slept   had
cooked   lived
drank

The Tale of
of
Brer Rabbit
and the
Tar Baby

BEFORE READING

 1 **Do you know these words?**

rabbit
兔子

fox
狐狸

bucket of tar
柏油桶

Tar Baby
柏油娃娃

buttons
鈕扣

bear
熊

briar patch
荊棘地

turtle
海龜

straw hat
草帽

comb
梳子

62

PART ONE

The Tar Baby

It was a hot day in August. Summer is a very hot season in the South of the United States. All the animals on the old plantation [1] had a rest.

Brer Fox was outside his house. He sat under a magnolia [2] tree and drank cold lemonade [3]. He was very hot. He was also angry and nervous.

Brer Fox didn't like Brer Rabbit. Before Brer Rabbit came to the old plantation, Brer Fox was a happy fox. The old plantation was a peaceful place. Brer Rabbit tricked [4] everyone. He tricked Brer Bear, Brer Turtle, Brer Wolf and Brer Fox. Brer Rabbit was a very intelligent rabbit. He was young and dressed well.

Brer Fox was tired of [5] Brer Rabbit. He decided to trick him. He went to his garden and took a big bucket of tar. He put other oils in the bucket too. Then he mixed the tar for a long time. The tar

---

1. **plantation** : 種植園。
2. **magnolia** : 木蘭。
3. **lemonade** : 檸檬水。
4. **tricked** : 欺騙。
5. **was tired of** : 感到厭倦的。

had a terrible odour [1]. It was very sticky [2].

Brer Fox went into his house. He went to the kitchen to look for an old straw hat, but he didn't find it. Then he went to the living room. He looked there too. Finally he went to the bedroom. In the bedroom he found an old straw hat, two buttons and a comb. He put them in a sack. Then he took the bucket of tar and walked to the road.

He threw the tar near a log [3] and made a big black Tar Baby. He put the old straw hat on the Tar Baby. He put on two buttons for the eyes. Then he put on the comb for the mouth. Brer Fox looked at his work and was happy.

The Tar Baby was ready! Brer Fox hid behind a big tree. He waited for Brer Rabbit to walk by. He waited and waited. It was very hot.

After an hour Brer Rabbit walked down the road. He was very happy. He walked, jumped and sang a song.

Brer Fox watched him from behind the tree. Suddenly Brer Rabbit saw the Tar Baby! He stopped and looked at it. Brer Rabbit was a friendly rabbit. He said, "Good morning! It's a hot day today!"

1. **odour** : 氣味。
2. **sticky** : 黏的。
3. **log** :

The Tar Baby smiled but didn't answer.

"I am from the old plantation," said Brer Rabbit. "Where are you from?"

The Tar Baby smiled but didn't answer.

Brer Fox watched everything from behind the tree. He wanted to laugh, but he didn't.

Brer Rabbit tried again. "Good morning! How are you?"

The Tar Baby smiled but didn't answer.

"What are you doing here? Where are you from?" Brer Rabbit asked again.

The Tar Baby smiled but didn't answer.

Brer Rabbit was angry. His face and ears were red.

"Can you hear me? I said 'Good morning!' Why don't you answer me?" he shouted.

There was no answer.

Brer Rabbit was very angry. "You are very unfriendly. I'm a friendly rabbit. I want to be your friend. Who are you?"

The Tar Baby didn't answer.

Brer Rabbit was furious [1]! He hit the Tar Baby! His front paw [2] was stuck in the Tar Baby's face [3].

Brer Fox was very happy. He laughed quietly.

"Let me go! [4]" said Brer Rabbit.

The Tar Baby did not let go.

---

1. **furious**：非常憤怒的。

2. **paw**：

3. **was stuck in the Tar Baby's face**：粘在柏油娃娃的臉上。

4. **Let me go!**：放開我！

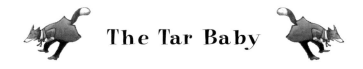

The Tar Baby

"Let me go!" said Brer Rabbit again.

Brer Rabbit kicked the Tar Baby. Now his back paw was stuck in the Tar Baby's body.

"Please, let me go!" shouted Brer Rabbit.

He kicked the Tar Baby again. Now his other back paw was stuck.

"Help! I can't move!" he shouted. "This is terrible!"

Poor Brer Rabbit! He was covered with tar. His face and ears were black with tar. His paws were black with tar. He was a very unhappy rabbit.

UNDERSTANDING THE TEXT

1 **Choose the correct answer.**

a. Brer Fox was angry and nervous because
☐ it was a cold day.
☐ he didn't like Brer Rabbit.
☐ Brer Bear tricked him.

b. Brer Rabbit was young and
☐ very funny.
☐ very peaceful.
☐ very intelligent.

c. Brer Fox got a bucket of tar because he wanted to
☐ trick Brer Turtle.
☐ play with Brer Rabbit.
☐ make a Tar Baby.

d. When Brer Rabbit saw the Tar Baby
☐ he ran away.
☐ he talked to it.
☐ he laughed at it.

e. Brer Rabbit got angry because
☐ the Tar Baby didn't answer him.
☐ the Tar Baby ran away.
☐ the Tar Baby laughed at him.

f. Brer Rabbit was covered with tar because
☐ he ate some tar.
☐ he hit and kicked the Tar Baby.
☐ he fell into the tar bucket.

2 **Odd one out!**

Circle the word that doesn't belong to the same category.

a. turtle   wolf   fox   paw   rabbit

b. intelligent   hungry   thirsty   sleepy   tired

c. bedroom   garden   kitchen   living room   bathroom

d. summer   winter   spring   August   autumn

e. ears   face   nose   mouth   baby

3 **Verbs that can trick you!**

Find the Past Simple（過去時）of these verbs in Part One. Underline them in red. Then write the Past Simple next to the Infinitive（動詞原形）.

Infinitive	Past Simple
sit	....
drink	....
come	....
trick	....
go	....
take	....
put	....
find	....
throw	....
make	....
sing	....
see	....
try	....

4 **Let's visit Brer Fox Manor!**

Label Brer Fox Manor. You'll find the words in the bucket.

garden
bathroom
living room
bedroom
kitchen

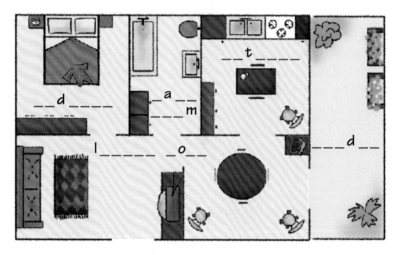

__ t __ __ __ __ __

__ __ d __ __ __ __

a __ __

__ __ __ m

__ __ __ d __ __

l __ __ __ __ __ __

o __ __

Now describe what Brer Fox does in these rooms. You'll find the words in the log.

sleeps   watches TV

eats   has a bath

a. Brer Fox .... in the kitchen.

b. He .... in the bedroom.

c. He .... in the living room.

d. He .... in the bathroom.

5 **Your home**

Circle the correct word or give the correct information.

Number of rooms: ....

Garden: ☐ Yes ☐ No

Living room: ☐ Big
☐ Medium
☐ Small

Your room: ☐ Big
☐ Medium
☐ Small

Number of bathrooms: ....

Your favourite room: ....

6 **Listen to the text and circle the words that are mentioned in each room.**

Let's look at Brer Fox Manor. First let's go to the living room. In the living room there are *books / beds* and there are two *tables / chairs*.

Now let's go to the kitchen. In the kitchen there's a *TV / table* and a *chair / clock*. There's a *plate / cup* on the table.

Now let's go to the bedroom. In the bedroom there's a *bed / bath*, a *fridge / clock* and *jeans / a jacket*.

Fox Hunting

In Great Britain fox hunting is permitted[1]. But many British people think it is cruel[2] to hunt foxes. They want a law to stop fox hunting.

In some states of the U.S.A. fox hunting is still permitted.

There are different types of foxes. The most popular is the red

A kit fox

fox. It lives in Europe. The San Joaquin kit fox lives in the North American deserts. In the Arctic we find the white Arctic fox. The Fennec fox is very small and it has big ears.

A Fennec fox

Today people want to help and protect animals, especially young animals. The WWF (World Wildlife Fund[3]) defends animals all over the world.

A red fox

An Arctic fox

1. **permitted**：被准許的。
2. **cruel**：殘忍的。
3. **WWF**：世界野生動物基金會。

1 Fill in the gaps with the words in the fox.

a. In Great Britain .... is permitted.

b. Many British people think it is .... to hunt foxes.

c. The most popular fox is the ....

d. The San Joaquin kit fox lives in the North American ....

e. The .... is very small.

f. The .... defends animals all over the world.

2 Think about this.

a. Why is it important to protect animals?

b. Do you know any organisations that protect animals?

c. Hunters and fires are dangers for animals. Can you think of any other dangers?

PART TWO

The Briar Patch

W hen Brer Fox saw Brer Rabbit covered with tar, he laughed and laughed. He walked down the road and said, "This is a good lesson for you, Brer Rabbit. You always tricked everyone on the old plantation. This time I tricked YOU! This is the end of Brer Rabbit!" Brer Fox looked at Brer Rabbit and laughed again.

Brer Rabbit didn't say one word. He was frightened[1]. He didn't move.

Brer Fox looked at his watch and said, "It's dinner time and I'm very hungry. I want rabbit barbecue for dinner. Rabbit barbecue is delicious. I must find some wood to make a fire." Brer Fox went to look for some wood.

Brer Rabbit started to think. He was a very intelligent rabbit. His eyes moved from left to right. He looked everywhere. Then he saw a briar patch. "The briar patch can take off[2] the tar. But I

---

1. **frightened** : 害怕的。　　　　2. **take off** : 除掉。

The Briar Patch

can't move. I'm stuck. I must go to the briar patch. What can I do?" thought Brer Rabbit.

Brer Fox returned and said, "I didn't find any wood to make a fire. I can't have rabbit barbecue for dinner, but I can hang[1] you!"

"Oh, Brer Fox, you can hang me, but please don't throw me in the briar patch!" said Brer Rabbit.

Brer Fox looked for a rope. "There's no rope. I can't hang you. How can I kill you?" asked Brer Fox.

He thought for a moment and said, "I can throw you in a river or a lake."

"Oh, Brer Fox, throw me in a river or a lake, but please don't throw me in the terrible briar patch. PLEASE!"

Brer Fox went to look for a river or a lake. He

1. **hang** :

looked everywhere but he didn't find a river or a lake. He was angry. He wanted to kill Brer Rabbit, but how?

"Have you got a heart [1], Brer Fox? Please don't throw me into the briar patch," said Brer Rabbit. "Oh, please!"

"Well," said Brer Fox smiling, "you don't want to go into the briar patch. That's exactly where I will throw you. Into the briar patch!"

Brer Fox threw Brer Rabbit into the briar patch! This was exactly what Brer Rabbit wanted. The Tar Baby stuck to the briar patch and Brer Rabbit was free!

When Brer Fox saw the Tar Baby in the briar patch he asked, "What's happening? Where is that cunning [2] rabbit? Why is the Tar Baby here?"

Brer Rabbit ran up the road and then stopped. He looked at Brer Fox and said, "You didn't listen to me. I said, 'Please don't throw me into the briar patch.' Next time you'll listen to me!" Brer Rabbit laughed and ran home to have a bath.

"You horrible [3] cunning rabbit. You tricked me again!" shouted Brer Fox. He was purple with anger. He looked at the briar patch and he looked at the Tar Baby. Then he walked home slowly. He was very sad and angry. Brer Rabbit tricked him again. Why was Brer Rabbit so intelligent?

When Brer Fox went into his garden he sat down under the magnolia tree. He was tired. He looked at the evening sky. He saw the stars and the moon and fell asleep.

---

1. **heart** :
2. **cunning** : 狡猾的。
3. **horrible** : 非常壞的。

UNDERSTANDING THE TEXT

1 ▸ **Fill in the gaps with the correct words from the comb. Some words can be used more than once.**

tar   hungry   briar patch

Tar Baby   dinner   wood   laughed

free   tricked

a. When Brer Fox saw Brer Rabbit covered with ....,
he ....

b. Brer Fox said, "I'm very .... I want rabbit
barbecue for ...."

c. Brer Fox went to look for some .... to make a fire.
He didn't find it.

d. Brer Rabbit said, "Oh, Brer Fox, please don't throw me into
the ...."

e. Brer Fox threw Brer Rabbit into the ....

f. The .... stuck to the briar patch and Brer Rabbit
was ....

g. Brer Fox was very angry and said, "You cunning rabbit. You
.... me again!"

2 **A word river**

How many animal names can you find in the word river? Circle them.

How many colours can you find? Underline them.
Use some of the words in the word river to complete these sentences.

a. Brer .... did not like Brer Rabbit.

b. Brer ...., Brer .... and Brer ....
also lived on the old plantation.

c. Brer Rabbit was .... from the tar.

d. At the end of the story Brer Fox was .... with anger.

e. Brer Fox sat under the .... tree.

3 **Match the following opposites. One is done for you.**

friendly   near

outside   put on

under   front

happy   over

young   inside

far   unfriendly

back   old

cry   laugh

take off   sad

Have fun with this crossword puzzle!

Across

1.

2.

3.

4.

Down

5.

6.

7. act dishonestly

8.

9.

10. black sticky substance

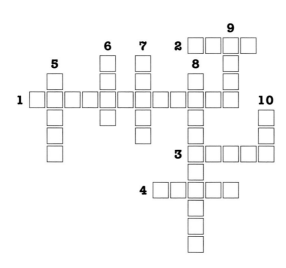

5 **Spidergram**

What other animal names do you know? Write them here. Two are done for you.

bear

turtle

Do you like animals? Why or why not?

What is your favourite animal? Why?

Do you have a pet at home? Describe it.

BRER TALES
and the South

Welcome to the South and welcome to the old plantation! The people of the South are famous for their hospitality [1]. Brer Tales were told about 200 years ago on the plantations of the South. The children of the plantations listened to these tales about funny animal characters. Everyone loved Brer Tales. In 1880 the American writer Joel Chandler Harris published these tales. His book was called *Uncle Remus: His Songs and Sayings.* Harris created a character named Uncle Remus. Uncle Remus told

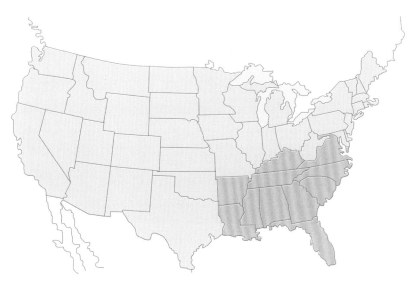

The Southern Plantation States

1. **hospitality** : 好客。

the tales and everyone liked him. There is also a film about Uncle Remus and his tales.

In the 1700's and 1800's there were many plantations in the South. Tobacco, cotton and sugar cane [1] were the most important products of the plantations. Every plantation had a big, beautiful house and garden. The owners of the plantation lived there. Today in the South of the United States you can visit old plantation houses.

Orton Plantation, North Carolina

1. **sugar cane**：甘蔗。

1 **Are these sentences true (T) or false (F)? Correct the false sentences.**

	T	F

a. Brer Tales were told about 200 years ago on the plantations of the South. ☐ ☐

b. The tales were about funny children. ☐ ☐

c. In 1880 the American writer Uncle Remus published the tales. ☐ ☐

d. Everyone liked Uncle Remus. ☐ ☐

e. In the 1700's and 1800's there were only two plantations in the South. ☐ ☐

f. The most important products of the plantations were tobacco, cotton and sugar cane. ☐ ☐

g. A plantation house was big and beautiful. ☐ ☐

Plantation gardens

EXIT TEST

CONTEXT

 Circle the correct word.

America became a nation ¹*before/after* the American Revolution. The enormous American continent ²*wasn't/was* a wilderness. American Indians lived here. Thousands of ³*settlers/soldiers* wanted to ⁴*live/fight* in these new lands. Life on the frontier was ⁵*easy/difficult*, but the settlers ⁶*died/were happy*. They loved the adventure and freedom of the ⁷*revolution/frontier*. The American Army ⁸*bought/built* forts in the wilderness.

In the West cowboys lived and worked ⁹*in a fort/on a ranch*. They moved ¹⁰*cattle/horses* to other places to ¹¹*sell/buy* them.

In the ¹²*South/North* there were ¹³*small/big* plantations. Tobacco, ¹⁴*cotton/cattle* and sugar cane were the most important products of the plantations.

COMPREHENSION

 Choose the words from the box to complete the sentences.

covered wagon   planted   cowboy   friend   library   lived   legend
King   coyote   ranch   frontier   West   Texas   animals
fell out   freedom   language

1. Johnny Appleseed .... apple seeds across America.

2. He was a .... of the Indians and the settlers.

3. He loved and respected all ....

4. He created the first .... on the frontier.

5. He made the .... a happy place.

6. Johnny Appleseed became a .... during his life.

7. Pecos Bill's family loved adventure and ....

8. They travelled to the .... in a ....

9. Little Bill .... of the wagon and a mother .... found him.

10. He .... with the coyotes and learned the animal ....

11. One day he met Tall Tom and decided to become a ....

12. Pecos Bill worked on a .... and became the .... of the Cowboys of ....

3 Are the following sentences true (T) or false (F)? Correct the false sentences.

	T	F
a. Brer Fox did not like Brer Rabbit because he tricked everyone.	☐	☐
b. One day Brer Fox took a big cup of tar and made a Tar Rabbit.	☐	☐
c. When Brer Rabbit saw the Tar Baby he talked to him, but the Tar Baby did not answer.	☐	☐
d. Brer Rabbit was very happy.	☐	☐
e. He kicked the Tar Baby and his paws were stuck in the tar.	☐	☐
f. When Brer Fox saw Brer Rabbit covered with tar, he helped him.	☐	☐
g. Brer Rabbit said, "Please throw me into the briar patch!"	☐	☐
h. Briar Fox threw him into the briar patch. The Tar Baby stuck to the briar patch and Brer Rabbit was free!	☐	☐
i. Brer Fox was very angry because Brer Rabbit tricked him again.	☐	☐

GRAMMAR

4 Put the verbs into the Past Simple tense.

1. An apple pie is (*make*) .... with apples.

2. Settlers (*travel*) .... in covered wagons.

3. The man (*build*) .... log cabins and (*hunt*) .... for food.

4. Cowboys (*sleep*) .... under the stars.

5. They (*stay*) .... in town for a few days and (*have*) .... a bath!

6. Uncle Remus (*tell*) .... the tales and everyone (*like*) .... them.

5 **Match the opposites.**

difficult	friend
old	cold
happy	sad
enemy	short
hot	easy
long	inside
outside	quickly
slowly	new

6 **Tick the correct answer.**

1. Tobacco, sugar and cotton were the most important products
- [ ] of the West.
- [ ] of the plantations.

2. The people in the South are famous
- [ ] because they are rich.
- [ ] for their hospitality.

3. Today there are many cattle ranches
- [ ] in the Southern states.
- [ ] in the Western states.

7 **Which is your favourite story?** ....

Why? ....

....

....

....

American Folk Tales

KEY TO THE EXERCISES AND EXIT TEST

THE LEGEND OF JOHNNY APPLESEED

Page 10 Exercise 2
a. sack
b. settlers
c. library
d. injure
e. messenger

Page 11 Exercise 3
Johnny walked to the west.

PART ONE

Page 16 Exercise 1
a. Johnny worked as a missionary with the Indians
b. "Go and plant apple seeds across America"
c. apple seeds
d. the Midwest
e. around the fields
f. apple orchards
g. the first library on the frontier

Page 17 Exercise 2
a. unkind
b. unable
c. unhappy
d. unhealthy
e. untidy
f. unknown

Page 18 Exercise 3
b. small – big
c. fat – thin
d. short – long
e. new – old
f. bad – good
g. last – first

Page 18 Exercise 4
thin, long, old, first

Page 18 Exercise 5
taught, had, said, took, bought, wore, gave, began, built, found, ate, told

Page 19 Exercise 6

PART TWO

Page 28 Exercise 1
thousands, everyone, apple seeds, village, message, fort, home, attack, forest, animal, happy, legend

Page 29 Exercise 2
a. some
b. any
c. any
d. some
e. some
f. any
g. any

Page 30 Exercise 3

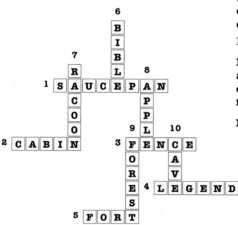

Page 31 Exercise 4
brave, settlers, name, enemy, took, fort, soldiers, save, friend, animals, legend

YOUNG AMERICA AND ITS SETTLERS

Page 34 Exercise 1
a. after the American Revolution
b. in small parts of the wilderness
c. to protect settlers and sell supplies
d. travelled in covered wagons
e. built log cabins and hunted for food
f. but the settlers loved the freedom and adventure

Page 36 Exercise 3
born, English, after, American, forests, deserts, small, forts, sold, log

PECOS BILL
PART ONE

Page 43 Exercise 1
a. East   **b.** adventure, West
c. Texas, covered wagon   **d.** coyote
e. twenty, cowboy, Tall Tom
f. cowboy, ranch

Page 44 Exercise 2
a. cactus
b. river
c. coyote
d. forest

Bill didn't have a tail.

Page 45 Exercise 3
a. was, decided   **b.** put, began
c. fell   **d.** went, slept   **e.** forgot
f. rode   **g.** walked.

Page 46 Exercise 4

 It's raining

 It's sunny

 It's windy

 It's foggy

 It's snowing

Page 46 Exercise 5
born, lot, father, West, wagon, forests, saw, met

PART TWO

Page 51 Exercise 1
a. they sat down to rest
b. were happy to meet Pecos Bill
c. Bucking Bronco
d. from Texas to other states
e. the King of the Cowboys of Texas
f. Big Bob

Page 52 Exercise 2
Pecos Bill's family wanted to move to the West.
The family's dream was to go to California.
During the journey, little Bill fell out of the family's covered wagaon.
Pecos Bill's horse is called Bucking Bronco. Pecos Bill's best friend is Tall Tom. Tall Tom's favourite food is apple pie.
Sue's home is near the Rio Grande River.
Pecos Bill's wedding celebration was exciting.

Page 53 Exercise 3

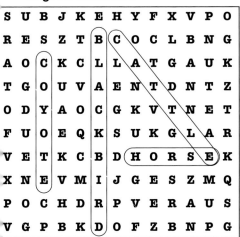

a. coyote
b. blackbird
c. cattle
d. horse

Page 54 Exercise 4
a1 - b3 - c6 - d5 - e4 - f2.

Page 54 Exercise 5
a. at
b. long
c. cattle
d. language
e. celebrated

Page 55 Exercise 6

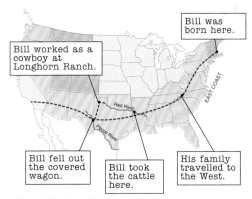

Bill was born here.

Bill worked as a cowboy at Longhorn Ranch.

Bill fell out the covered wagon.

Bill took the cattle here.

His family travelled to the West.

LIFE ON A RANCH

Page 59 Exercise 1
a. F – A ranch is a very big piece of land.
b. F – The cowboys lived in a bunkhouse.
c. T
d. F – The cowboys' work was difficult.
e. T
f. F – They stayed in a town for a few days.
g. T

Page 60 Exercise 2
<u>At the Ranch</u>
a. worked, lived
b. ate

<u>At the Saloon</u>
a. played
b. drank, had

<u>On the Plains and in the Mountains</u>
a. cooked
b. slept

THE TALE OF BRER RABBIT AND THE TAR BABY

PART ONE

Page 68 Exercise 1
a. he didn't like Brer Rabbit
b. very intelligent
c. make a Tar Baby
d. he talked to it
e. the Tar Baby didn't answer him
f. he hit and kicked the Tar Baby

Page 69 Exercise 2
a. paw
b. intelligent
c. garden
d. August
e. baby

Page 69 Exercise 3
sat, drank, came, tricked, went, took, put, found, threw, made, sang, saw, tried

Page 70 Exercise 4

a. eats
b. sleeps
c. watches TV
d. has a bath

Page 71 Exercise 6
books, chairs, table, chair, cup, bed, clock, a jacket

FOX HUNTING

PART ONE

Page 73 Exercise 1
a. fox hunting
b. cruel
c. red fox
d. deserts
e. Fennec fox
f. WWF

PART TWO

Page 77 Exercise 1
a. tar, laughed
b. hungry, dinner
c. wood
d. briar patch
e. briar patch
f. Tar Baby, free
g. tricked

Page 78 Exercise 2
Seven animal names. Four colours.

a. Fox
b. Turtle, Bear, Wolf
c. black
d. purple
e. green

Page 78 Exercise 3
outside – inside
under – over
happy – sad
young – old
far – near
back – front
cry – laugh
take off – put on

Page 79 Exercise 4

```
            6    7    2  ┌─┬─┬─┬─┐   9
                        │F│I│R│E│
       5   ┌─┐  ┌─┐  8  ├─┼─┴─┴─┤
          │M│  │T│     │O│
       ┌─┐├─┤  ├─┤  ┌─┐├─┤
       │W││O│  │R│  │B││P│
   ┌─┬─┼─┼─┼─┬─┼─┼─┬─┼─┼─┼─┐      10
 1 │M│A│G│N│O│L│I│A│T│R│E│E│
   ├─┤├─┤  ├─┤  ├─┤      ┌─┐
   │T││O│  │N│  │C│      │T│
   ├─┤├─┤  └─┘  ├─┤      ├─┤
   │C││N│       │K│      │A│
   ├─┤└─┘         ┌─┬─┬─┬─┼─┤
   │H│          3 │R│I│V│E│R│
   └─┘            └─┴─┴─┴─┴─┘
                    ┌─┐
                    │P│
              ┌─┬─┬─┼─┼─┐
            4 │S│T│A│R│S│
              └─┴─┴─┼─┼─┘
                    │T│
                    ├─┤
                    │C│
                    ├─┤
                    │H│
                    └─┘
```

Page 80 Exercise 5
Open answer

BRER TALES AND THE SOUTH

Page 83 Exercise 1
a. T
b. F – The tales were about funny animal characters.
c. F – in 1880 the American writer Joel Chandler Harris published the tales.
d. T
e. F – In the 1700's and 1800's there were many plantations in the South.
f. T
g. T

KEY TO EXIT TEST

1. **1.** after **2.** was **3.** settlers **4.** live **5.** difficult **6.** were happy
7. frontier **8.** built **9.** on a ranch **10.** cattle **11.** sell **12.** South
13. big **14.** cotton

2. **1.** planted **2.** friend **3.** animals **4.** library **5.** frontier **6.** legend
7. freedom **8.** West / covered wagon **9.** fell out / coyote
10. lived / language **11.** cowboy **12.** ranch / King / Texas

3. **a.** T
b. F – He made a Tar Baby.
c. T
d. F – He was very angry.
e. T
f. F – He didn't help him.
g. F – Brer Rabbit said, "Please don't throw me into the briar patch!"
h. T
i. T

4. **1.** made **2.** travelled **3.** built / hunted
4. slept **5.** stayed / had **6.** told / liked

5. difficult – easy
old – new
happy – sad
enemy – friend
hot – cold
long – short
outside – inside
slowly – quickly

6. **1.** of the plantations
2. for their hospitality
3. in the Western states

7. Open answer.

91

N🍎TES

N⦿TES

N🍎TES

Black Cat English Readers

BLACK CAT ENGLISH CLUB
Membership Application Form

BLACK CAT ENGLISH CLUB is for those who love English reading and seek for better English to share and learn with fun together.

Benefits offered: - *Membership Card*

- *Member badge, poster, bookmark*

- *Book discount coupon*

- *Black Cat English Reward Scheme*

- *English learning e-forum*

- *Surprise gift and more...*

Simply fill out the application form below and fax it back to **2565 1113**.

Join Now! It's FREE exclusively for readers who have purchased *Black Cat English Readers* !

The book(or book set) that you have purchased: _____

English Name:_____ (Surname) _____ (Given Name)

Chinese Name: _____

Address: _____

Tel: _____ Fax: _____

Email:_____

Sex: ❏ Male   ❏ Female   (Login password for e-forum will be sent to this email address.)

Education Background: ❏ Primary 1-3   ❏ Primary 4-6   ❏ Junior Secondary Education (F1-3)

❏ Senior Secondary Education (F4-5)   ❏ Matriculation

❏ College   ❏ University or above

Age: ❏ 6 - 9   ❏ 10 - 12   ❏ 13 - 15   ❏ 16 - 18   ❏ 19 - 24   ❏ 25 - 34

❏ 35 - 44   ❏ 45 - 54   ❏ 55 or above

Occupation: ❏ Student   ❏ Teacher   ❏ White Collar   ❏ Blue Collar

❏ Professional   ❏ Manager   ❏ Business Owner   ❏ Housewife

❏ Others (please specify: _____)

As a member, what would you like **BLACK CAT ENGLISH CLUB** to offer:

❏ Member gathering/ party   ❏ English class with native teacher   ❏ English competition

❏ Newsletter   ❏ Online sharing   ❏ Book fair

❏ Book discount   ❏ Others (please specify: _____)

Other suggestions to **BLACK CAT ENGLISH CLUB**:

_____

Please sign here: _____

(Date:_____)